Creative Genius

Short Stories For Boys

Written by Shermaine Perry-Knights
Illustrated by Oleksandr Zinchenko

*Drew,
Every book is a new adventure!*

Copyright © 2023 by Innovation Consultants of DeKalb.
Written by Shermaine Perry-Knights.
Illustrated by Oleksandr Zinchenko.

eBook ISBN: 978-1-953518-36-1
Paperback ISBN: 978-1-953518-37-8
Hardcover ISBN: 978-1-953518-38-5

No part of this book may be used or reproduced in any manner whatsoever without the prior written permission of the author.

Table of Contents

Dedication.................................. 5
Short Stories For Boys 6
   Express Yourself 7
   Bounce Back........................15
   Beautiful Differences 23
   Peer Pressure 31
   Just Relax............................39
   Shine Bright 47
At Home Word Wall 55
Vocabulary List.......................60
Teacher Resource 62
Collect Them All 67
About the Author 68

Dedication

This book is dedicated to every boy that loves the thrill of adventure. You deserve to feel heard and seen in books. Books are the best kind of adventure. Enjoy and don't forget to tell a friend!

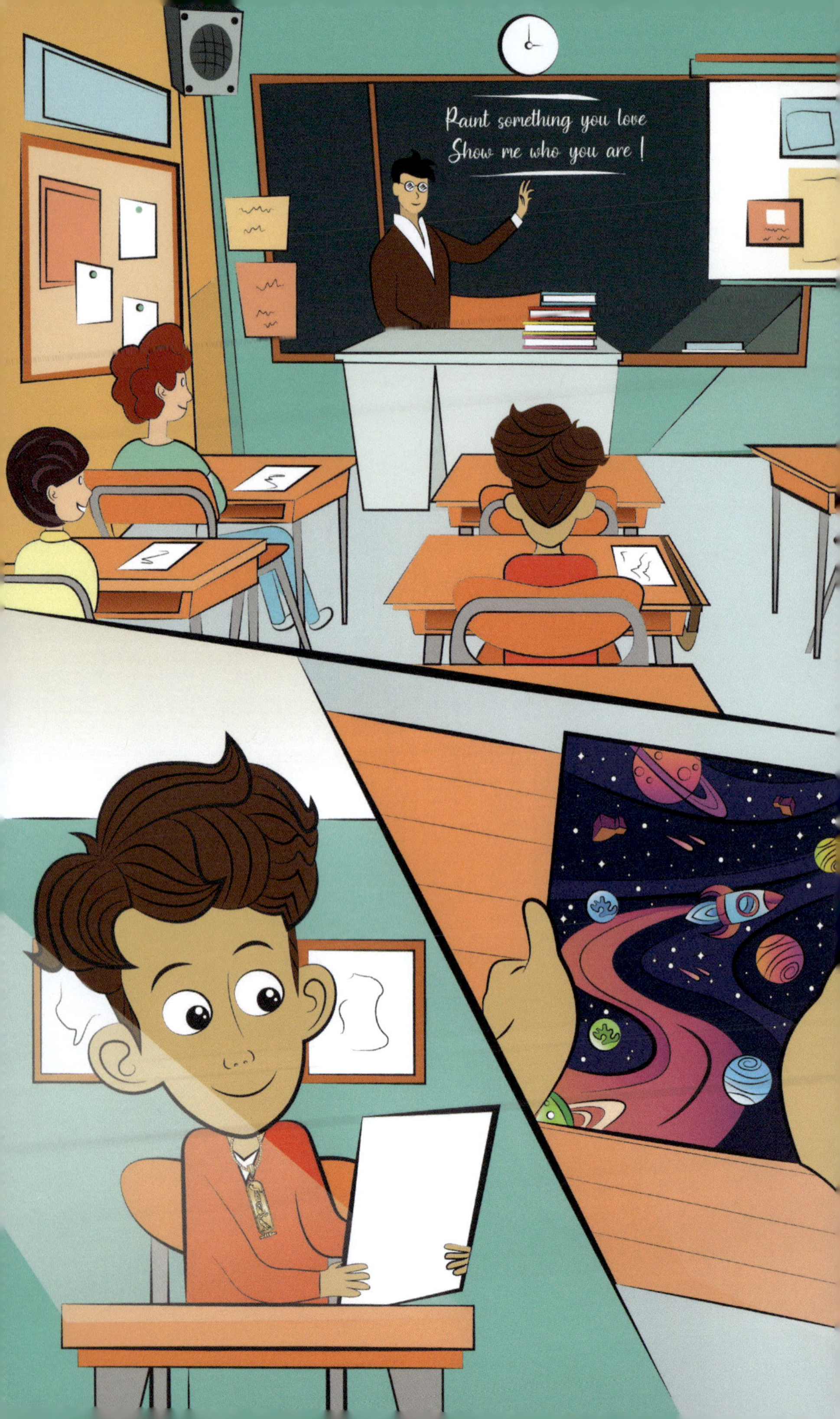
Paint something you love
Show me who you are !

Express Yourself

Tom hunched over his desk in art class. The room was brightly lit and smelled of old paint and crayons. He gripped his paintbrush and dipped it into the violet paint blobbed on his palette. He chewed his lip, flicking his gaze between the paper laying on the table in front of him, and the instructions written on the board in whiteboard marker.

"Paint something you love. Show me who you are!" it read. The art teacher, who now sat at his desk next to the board smiled at Tom.

He looked back at his paper. An idea flashed into his mind. His paintbrush moved, dancing over the paper as the shapes and colors flooded Tom's mind. When he finished, he leaned back and marveled at his work. It was a dark purple spaceship zooming through a solar system of planets with strange patterns and colors.

He loved spaceships. Tom also loved Stars Wars. He enjoyed cosplay and planned to dress as Darth Vader for Halloween. Science fiction had a special place in his heart. One day, he hoped to write his own science fiction novel.

"You are a nerd," Damon said, peering at his painting as the bell rang and the students gathered their belongings. He had a friendly smirk on his face.

"What's wrong with that?" Tom asked with a shrug.

Damon pointed to his own painting, which was of a crimson red racing car. "Some people don't think science is cool."

Tom's face flushed. He hid his painting by facing it towards his body as he brought it to the teacher who hung it on the drying line.

"This is a lovely picture, Tom!" the teacher said, eyes glistening in the white light above his head. "Do you love space travel?"

"Um, I'm not sure," Tom said, before spinning around and fleeing the room.

Tom and Damon's next class was science.

"Today, we are going to learn about the sun and what it does for us," the science teacher said, flicking the PowerPoint on at the front of the class.

A few groans popped up from the students around the room. Tom, however, smiled. He found the sun fascinating and understood it provided life for humanity in various ways.

"Can anyone guess the distance of the sun to earth?" Mr. Cromwell asked the class.

Tom's heart thumped in his chest. He knew the answer. His hand almost shot in the air. But he was afraid of not looking cool. He didn't want to look like a nerd to everyone else. So, his answer of 149.6 million kilometers

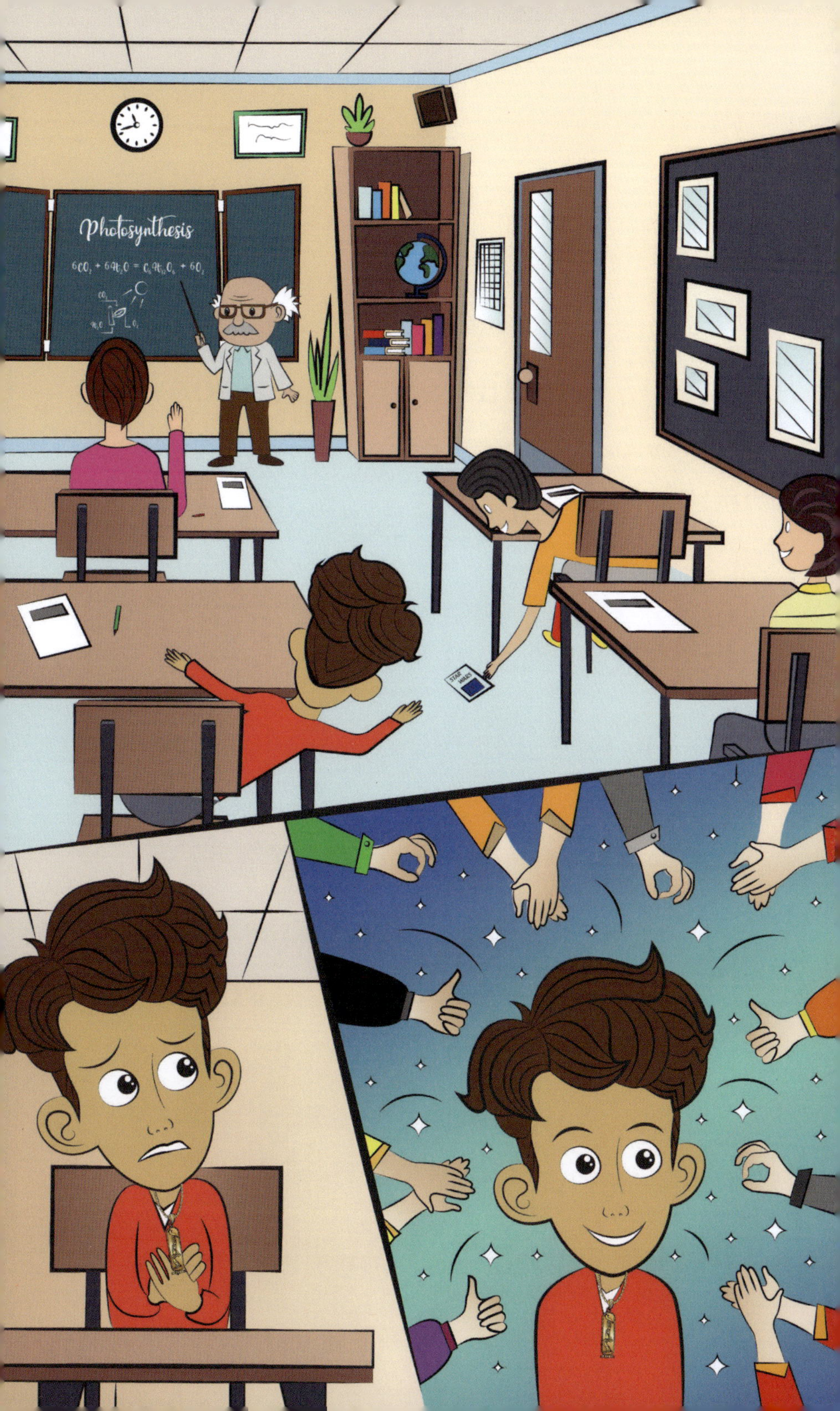
Photosynthesis

died on his tongue. He chanced a glance around the room at the other students who appeared disinterested.

"Plants and trees absorb the sun's light to keep them alive. Does anyone know what this process is called?" Mr. Cromwell continued, flicking through his slideshow.

I do! Tom thought. *It's photosynthesis!*

But he kept his mouth zipped shut. The teacher asked them all to grab out their workbooks and do an activity. Tom reached for his backpack. As he opened it, a flyer to a local Star Wars virtual reality immersive experience slipped out, sliding across the floor, underneath Damon's desk.

He scrunched his brows and plucked it

up. He read it aloud.

Tom heaved a sigh, burying his head in his hands as his face went as red as a tomato.

"What? This is so cool!" Damon blurted out. "I want to go!"

Tom stared at Damon, wide-eyed as the class erupted into whoops and cheers.

"We should plan to go!" a girl said.

"I agree!" Damon said. He patted Tom on the back. "Thanks for bringing this!"

Tom smiled, and he couldn't believe it. Other kids loved space travel and Star Wars just like him!

The End.

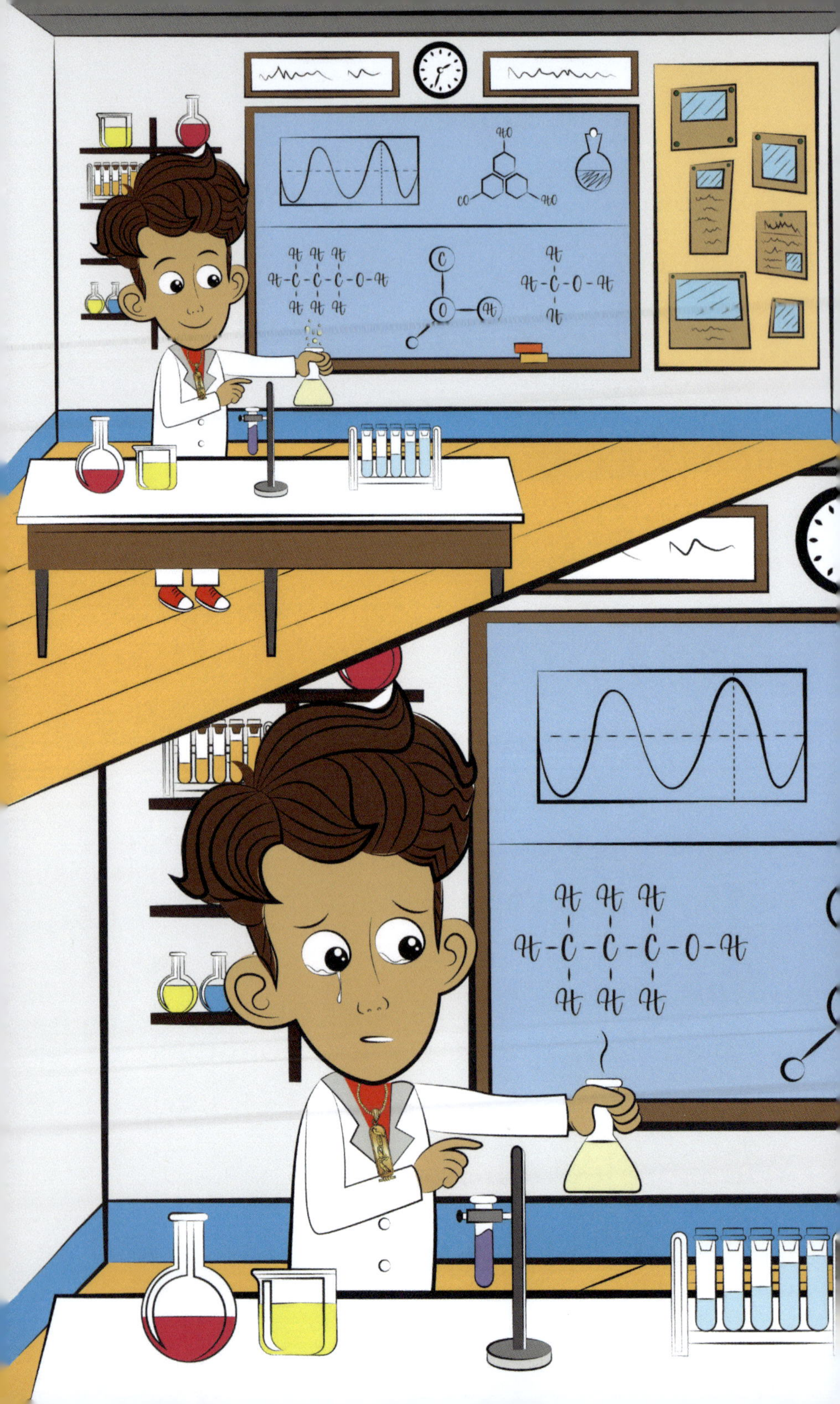

Bounce Back

Tom woke up on a Friday morning with a grin. He threw himself out of bed, dressed, and practically hurled himself onto the bus. He couldn't wait to get to school for the science experiment his class was doing.

"Elephant toothpaste," Mr. Cromwell announced when he scurried into the science classroom later that morning.

Everyone laughed. "Where is the elephant?" Damon asked, looking around. He made a show of peeping under his desk and under his bag.

Mr. Cromwell shook his head, hands on his hips. "I'm afraid there are no elephants for us to look at. However, if we all perform this experiment safely, we can go on a class trip to the zoo to see the elephants."

Cheers and whoops erupted from the class like an active volcano. "The elephant toothpaste experiment is an advanced test. Other teachers would likely choose to simply demonstrate it to you themselves. But I believe you are all responsible enough to do it yourselves. Come up and grab your supplies and the instructions card."

Tom shouldered his way to the front of the class and grabbed what he needed. He began the experiment, pulling his protective glasses onto his face, and arranging the test tube on the rack. After a quick scan of the instructions, he poured in the dish soap, purple food dye, and hydrogen peroxide. He stared at the test tube, waiting for the chemical reaction. But nothing happened. He did not know he had missed the final step.

He glanced around the room as everyone else's elephant toothpaste shot from their test tubes in colorful foam.

A bubble of frustration built up in his chest. He swallowed the tears prickling his eyes. As he turned around to ask Damon what he'd done wrong, he accidentally knocked his test tube, spilling the contents out. It gushed across his desk and splashed his hands.

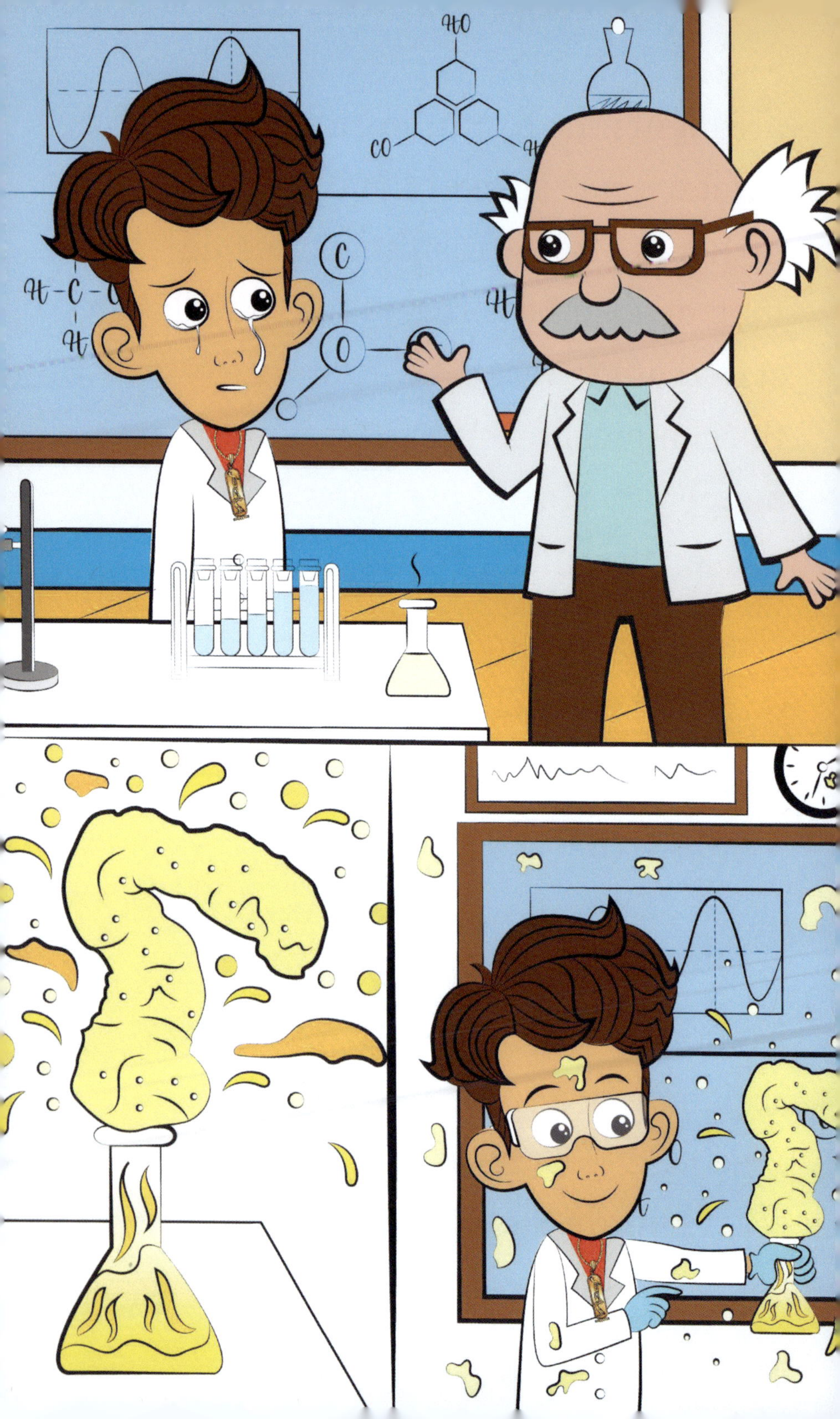
HO
CO
C
O
H-C-C
H

"Ouch!" he gasped, jerking his hand away. The hydrogen peroxide, which is a chemical, stung his bare skin.

"Tom, why are you not wearing gloves?" Mr. Cromwell said, standing over him with a deep frown.

Tom's stomach wound into knots. "I... I didn't mean to. I forgot."

"And now you are hurt," Mr. Cromwell said. Tears spilled down Tom's cheeks as Mr. Cromwell led him to the sink to rinse his stinging hand in cold water.

"I'm sorry, sir," Tom said, voice wavering. "I'm going to miss out on the zoo, aren't I?"

Mr. Cromwell sighed, pressing his lips together. "Do you want to try the experiment again?" he whispered.

"But I failed," he said, eyes watery.

"We all make mistakes sometimes," Mr. Cromwell said. "Failures and mistakes are another opportunity to learn and grow. You can learn from this failure by trying again and correcting your mistakes."

Tom stood for a moment, letting his disappointment wash through him like waves over a shore. He drew in a deep, shaky breath and wiped his tears. "Let's do it."

He trudged back to his station and read the instructions once more. "I forgot the yeast and hot water, the final step," he said to Mr. Cromwell. "That's why it didn't react."

Tom prepared his experiment the

same way he did the first time. But now, he wore protective gloves. He stirred the yeast and warm water before pouring it into the test tube with the other ingredients.

BAM! The elephant toothpaste, in its foamy, purple glory, oozed out of the test tube at break-neck speed. Tom grinned and laughed. "I did it!" he said.

"Good work, Tom," Mr. Cromwell said, clapping. "You have just mastered the art of bouncing back from disappointment!"

The End.

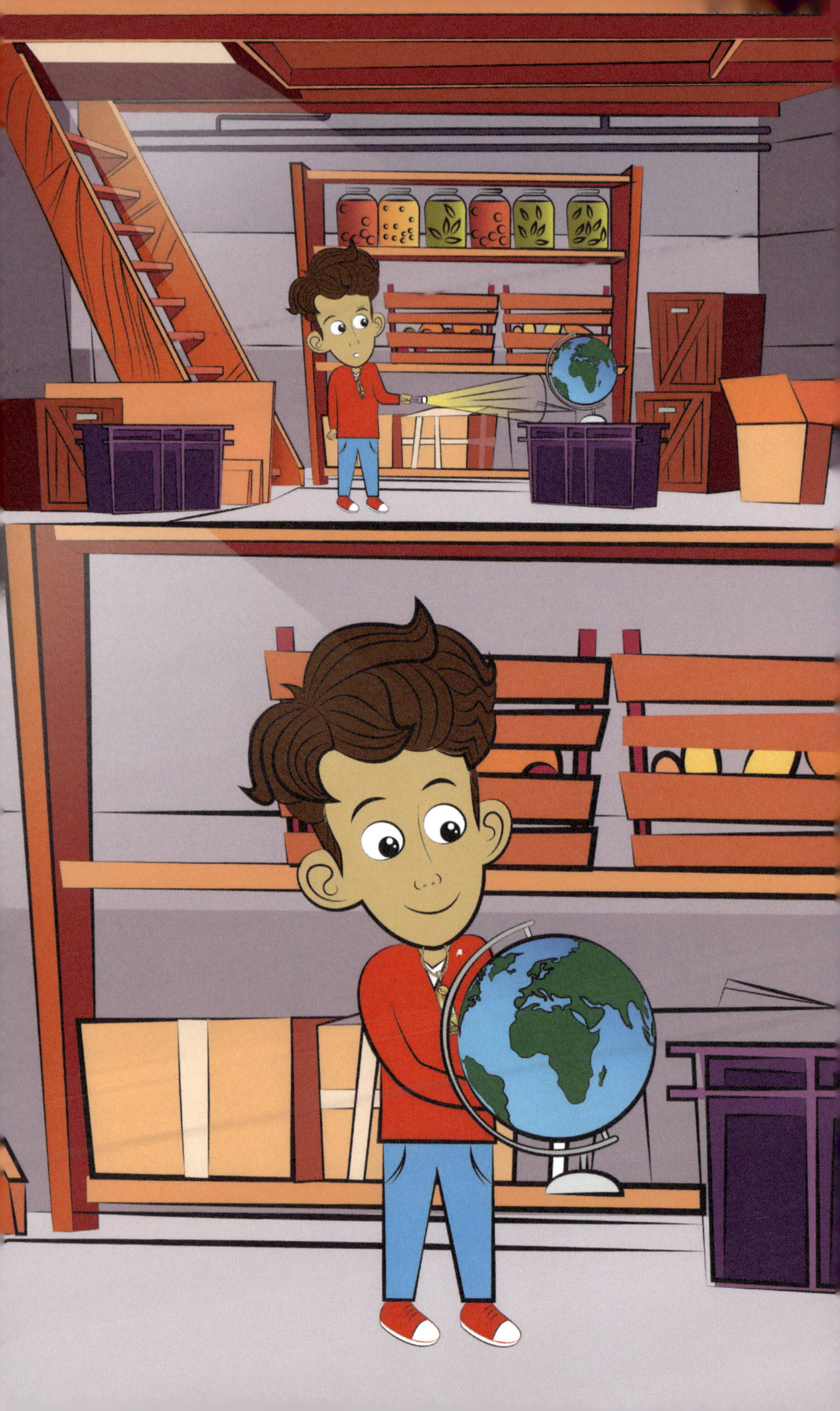

Beautiful Differences

The darkness of the basement made a shiver spider-walk down Tom's spine. He flicked on his flashlight, and his eyes widened. Dust bunnies hugged the corners of the basement. It smelled of mildew, and the floorboards creaked as he explored the room. As he roamed around the room, something caught his eye. It was a world globe. He grabbed from the shelf and took it to his room. In the light, he spun it slowly, noticing the different continents of the world, and the hundreds of countries. He realized there must be thousands of different cultures around the world.

The next day at school, his curiosity about different cultures still gnawed at his gut. He gathered his group of friends after school. They sat in a circle beneath the tree overlooking the football field.

Tom pulled out his globe. "I found this yesterday at home. I wanted to create a family tree but for friends to show where we are each from in the world."

"I'm Turkish," Anastasia said, brushing her brown curls behind her ears. She reached for the globe, and found Türkiye, pointing to it. "It's the only country in the world that is considered both in Europe and Asia.

Omar
Anastasia
Tom
Kim
Simon

Soccer is our most popular sport! My mother is from Istanbul."

"What's the best food from Türkiye?" Tom asked, curious.

"The Iskender Kebab, or Döner Kebab!" Anastasia said. "It's grilled over an open fire! We also love drinking tea. Did you know Türkiye has over 80,000 mosques? It's because so much of the population is Muslim."

"I'm from China," Kim said, showing the group on the globe. "China has the largest population in the world of over 1.4 billion people! Vegetables and rice are the most common ingredients in Chinese food. But some people eat scorpions on a stick. My father loves soccer too. My mother is from Beijing and every Chinese New Year, we eat dumplings. The Chinese New Year celebration lasts for 15 days. One of twelve animals represent each year."

"Scorpions on a stick!" Tom scrunched his face. "Brave people."

"We also consider China one of the four ancient civilizations," Kim said, grinning with pride at her culture. "China's history is 3000 years old."

Kim passed the globe to Omar. He pointed to his country. "I'm from Saudi Arabia. It's in the Middle East. Camels are still a special part of our culture and diet. My country is still a monarchy, meaning we have a king. We also love soccer! Our culture is quite traditional, and they consider poetry the most important form of art. A famous dish from my country is called Kabsu. It's a combination of roast chicken and rice."

Finally, Omar handed the globe to Simon. "I'm from Trinidad, which is an island of the nation, Trinidad and Tobago. It's in the Caribbean. We have over 100 mammal species, including monkeys and bats. We are known for our variety of seafood dishes, including curried crab! We love soccer too! Cricket and soccer are our most popular sports. Our culture is rich with music and dancing! So many people visit Trinidad for Carnival. A celebration that is known as the Greatest Show on Earth. Others visit for the clear blue water. Snorkeling and diving are a lot of fun."

Tom finished drawing everyone on his friend tree. "Look at this," he said with a grin. "We have a lot of differences, but that's what makes each of our cultures unique and beautiful."

"Seems like we are all connected, especially through soccer," Anastasia said.

Omar grinned, taking the globe back and spinning it slowly. "Unique but connected because we are all human."

The End.

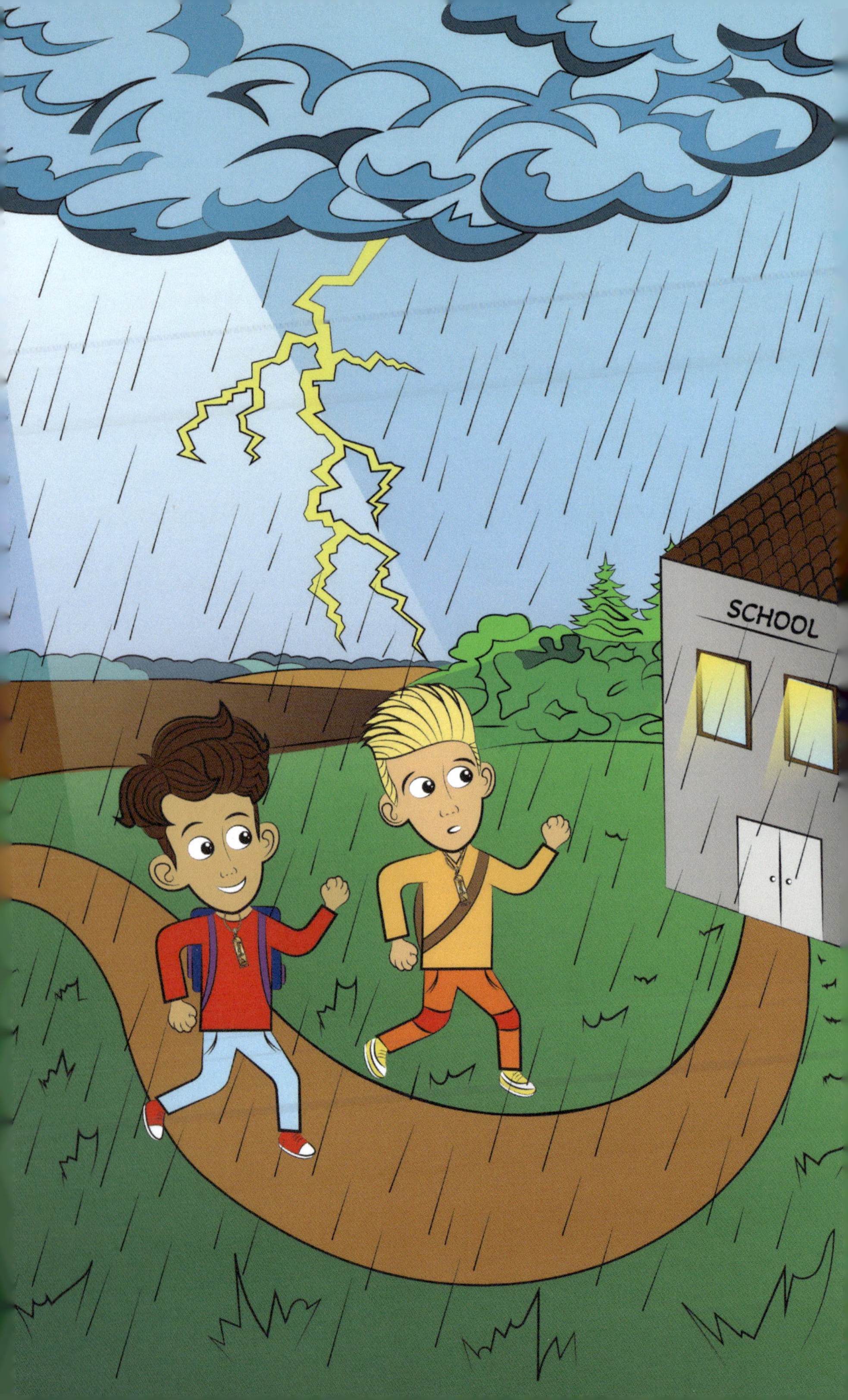
SCHOOL

Peer Pressure

Tom and Percy raced down the path from the school bus as thunder clapped above their heads and lightning shattered the sky. They shrieked with laughter as they slid into the school entrance, escaping the rain before it pelted down.

They scurried to their lockers and pulled out the textbook for science class. Tom bounced on the balls of his feet, giddy from the thunder and the excitement of the experiment they were conducting in class today.

"Good morning, students," Mr. Cromwell said. "What a great day for making a volcano!"

The class pinned their eyes on the teacher, listening as they were all excited about the lesson.

Mr. Cromwell gestured to his cart of supplies. "In pairs, we will each construct our own volcano with modelling clay. Then we will make it erupt with the fusion of baking soda and vinegar!"

Tom and Percy collected their tray of supplies. Tom grabbed the instructions card and glanced over the method.

"Oh, we don't need to read the instructions," Percy said, waving his hand in dismissal. He grabbed the blob of clay and began molding it into a cone shape.

"But we need to follow the instructions to know what to do," Tom said, furrowing his brow. He jabbed the instructions with his finger. "It says to put the mold on the clay around the test tube."

"Oh, I was just going to make the cone and then carve a hole for the tube," Percy said with a shrug.

Tom frowned as he spotted the blue food dye Percy had selected. "Volcanic lava isn't blue, Percy!"

Percy shook his head. "So? It's more fun with blue lava."

Tom chewed his lip, feeling a bubble of nerves at the conflict brewing

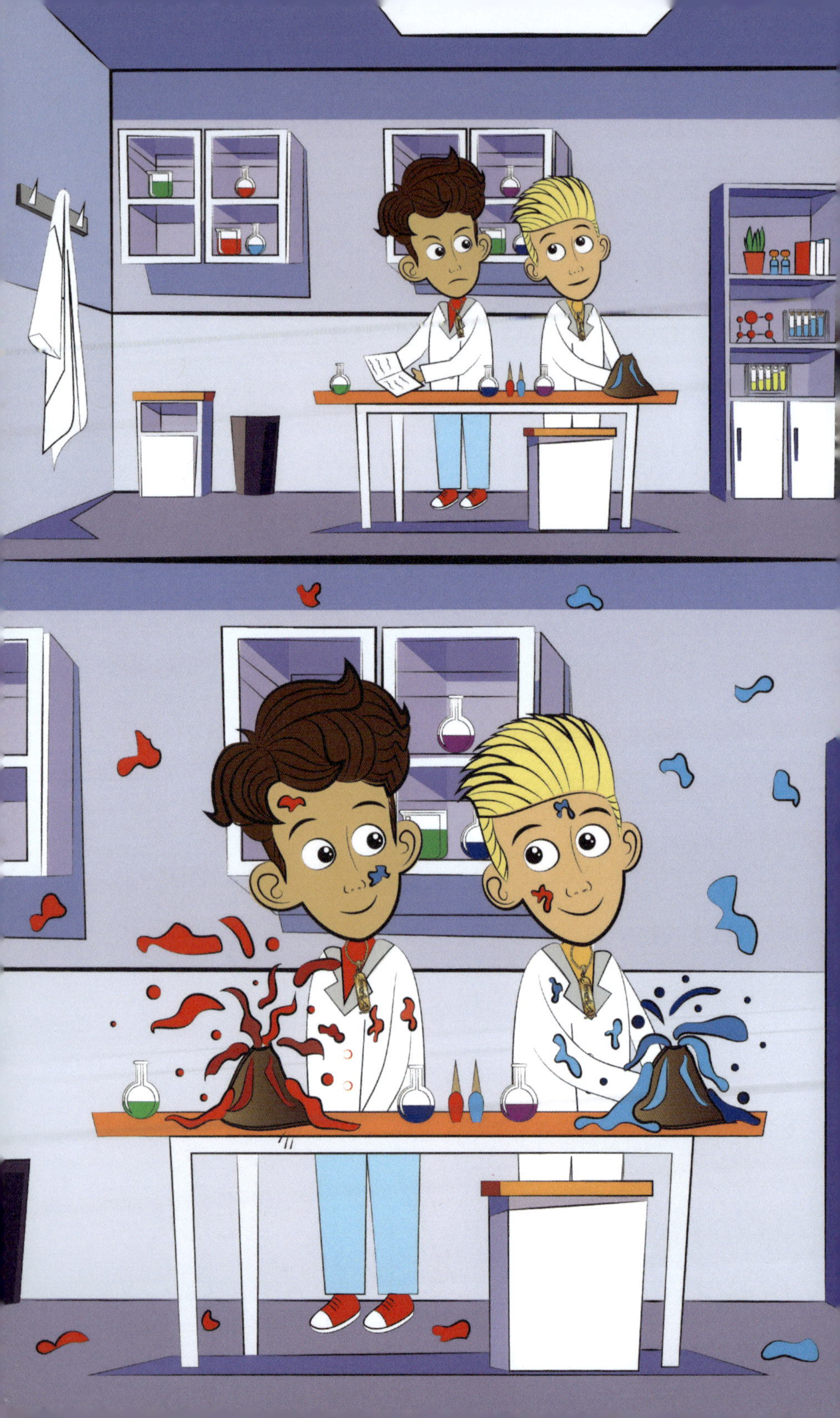

between them like a potion in a witch's cauldron. "But not scientifically accurate. We need red dye," he said.

Percy rolled his eyes. "We don't have to follow everything by the book, Tom. Besides, if I make a mistake, it's a great way for me to learn."

"I prefer learning by following the instructions. And I want red lava."

"But I want blue."

The boys frowned at one another. Then Percy heaved a sigh. He smashed his cone of clay and divided it into two equal blobs. Then he grabbed red food dye from the supply cart. He handed Tom a blob of clay and the red dye.

"You do things your way, and I'll do things my way," Percy said, offering his friend a smile. "Neither of us are wrong. We just have differences."

Tom beamed. "You are right. Should we have a competition to see who can make the best erupting volcano?"

Percy nodded, knitting his brows together in grim determination. "I'll beat you." He flashed his teeth in confidence.

The boys spent the rest of the lesson building their volcanos. Percy decided on the steps himself. While Tom followed the instructions.

The blue and red lava erupted from their volcanos as they each poured their vinegar into the test tube of baking soda.

"Damon!" Tom called their friend

over, who worked in the row in front of them. "Whose volcano is the coolest?"

Damon narrowed his eyes, studying each volcano. "I think mine is the best. Sorry!" he said with a chuckle.

Damon stepped aside, revealing his creation. His volcano oozed green lava!

"Ours are all different!" Tom said, marveling at Damon's work.

"They are all great," Mr. Cromwell said as he came by. "While they are all different, they are each special. Just like you!"

The End.

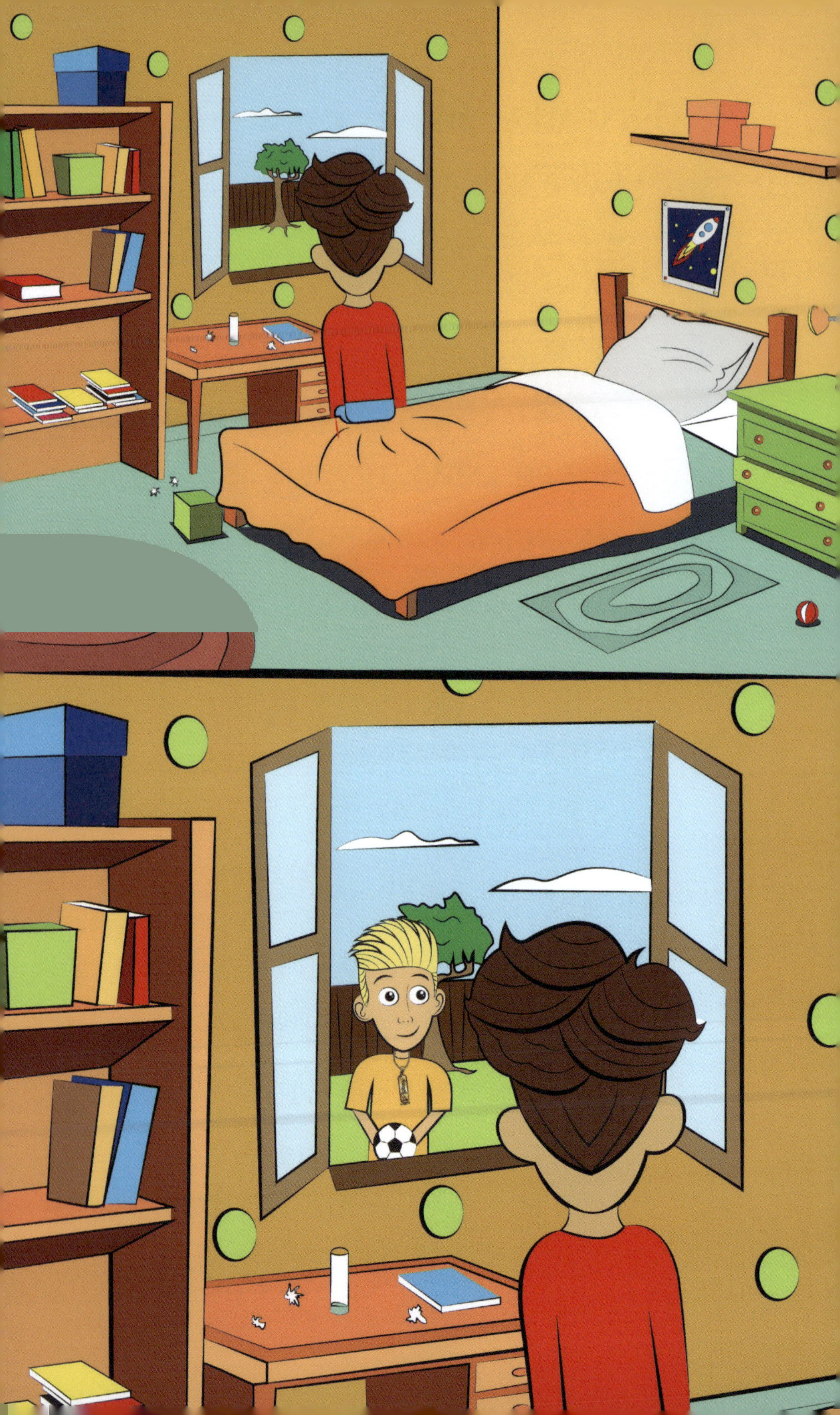

Just Relax

The evening sunlight glared through Tom's bedroom window one early Wednesday evening. His brows furrowed as he hunched over his desk. Textbooks laid open. His workbooks and notebooks are too. Pens and pencils and scrunched up paper scattered his desk. There was an empty glass where juice once was. He worked through his math equations, then moved onto his English readings.

There was a knock on his bedroom window, and he glanced up. It was Percy.

"Are you coming to the park, Tom?" Percy asked, holding up the soccer ball.

Tom shook his head, pressing his lips together. "I can't. I have so much homework to do. Then I must make a slideshow for the science club."

Perry frowned, lowering the ball. "Sounds like you're busy. I'll come back tomorrow."

The next day, Percy knocked on his window even though the wind howled, and the rain drummed the roof. "Can I come in? We could do some coloring. I have planet themed coloring books." He pointed to his backpack.

"Sorry, Percy!" Tom said, voice strained as he hardly cast his friend a glance. "I must practice my speech for French class, and I have to study for the pop quiz in science class tomorrow."

Percy slouched his shoulders. He traipsed away into the grey evening. Percy kept trying to drag Tom out of the house for fun activities for the next few days. But each time, Tom was busy. Too busy. His face was wild and had bags under his eyes from not enough rest. His hair was a mess. When he spoke to Percy, his words spilled from his mouth like water bursting from a dam.

Finally, Percy knew he had to step in. Instead of knocking on Tom's bedroom window, he knocked on the front door. Tom's mom let him in.

"What are you doing here?" Tom asked, jumping and throwing his paper into the air from fright.

Percy tried not to laugh at Tom's frazzled state. He heaved a sigh and sat on the edge of Tom's bed.

"Are you doing okay, Tom?" he asked.

Tom didn't even look at him as he returned to his books. "I'm fine. Why?"

"Tom, stop!" Percy said, voice sharp, clapping his hands. "You're acting like a robot."

Tom flinched and snapped his head to Percy. "A robot? Why?"

"Listen to me," Percy said, softening his expression and his voice. "Tom, I'm a little concerned you are working yourself too hard. Don't forget about the importance of relaxing."

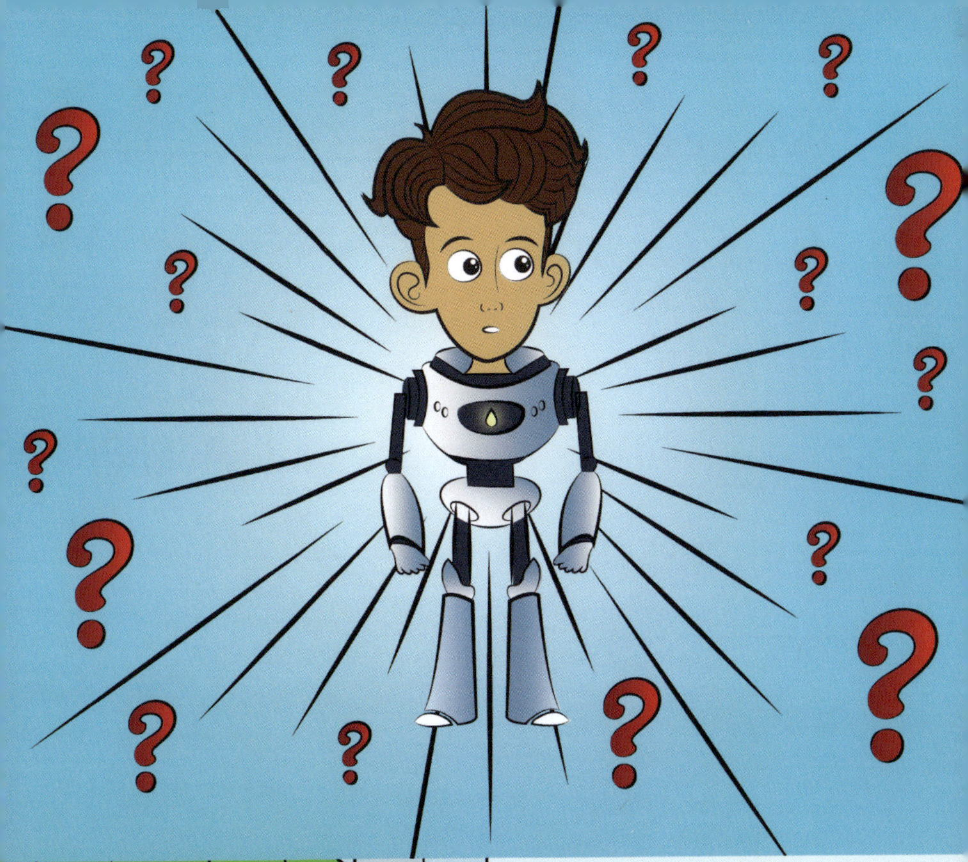

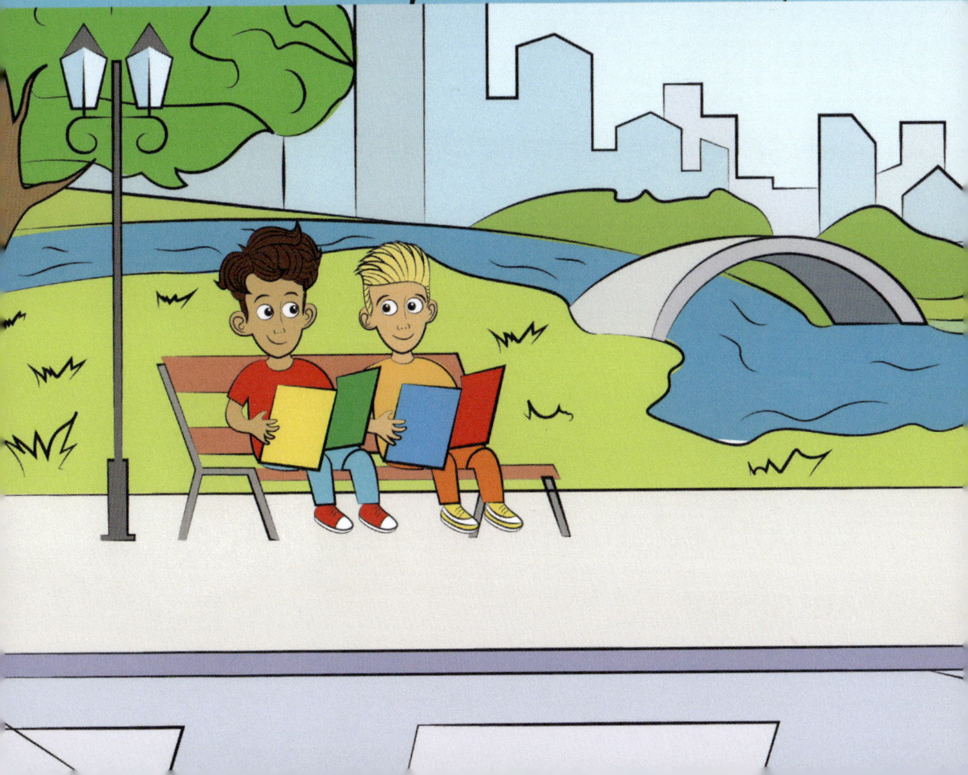

"Relaxing?" Tom said as if he'd never heard of the word.

"Yes," Percy said. "As growing kids, it's important we take a step back from our responsibilities every day and unwind."

Tom chewed his lip. "How?"

Percy thought about it. "You could turn your phone off and take an hour afternoon nap. Or color. Do cartwheels in the grass. Or a quick run through the neighborhood. Something that isn't schoolwork. Something you enjoy for yourself to release your stress."

Tom sat for a moment, taking this in. Then he shut his workbooks and nodded. "You're right, Percy. I've been exhausted. It feels like I'm running on autopilot!"

"So, do you want to come to the park with me?" Percy asked, digging into his backpack, pulling out some coloring books. "It's a great day for coloring by the river!"

Tom grinned. "Let's go!"

The two friends made their way to the park and found themselves a spot by the river where they colored. As the colors danced across the page, Tom felt his shoulders loosen and his heart slow. Percy was right. It was so important to unwind!

The End.

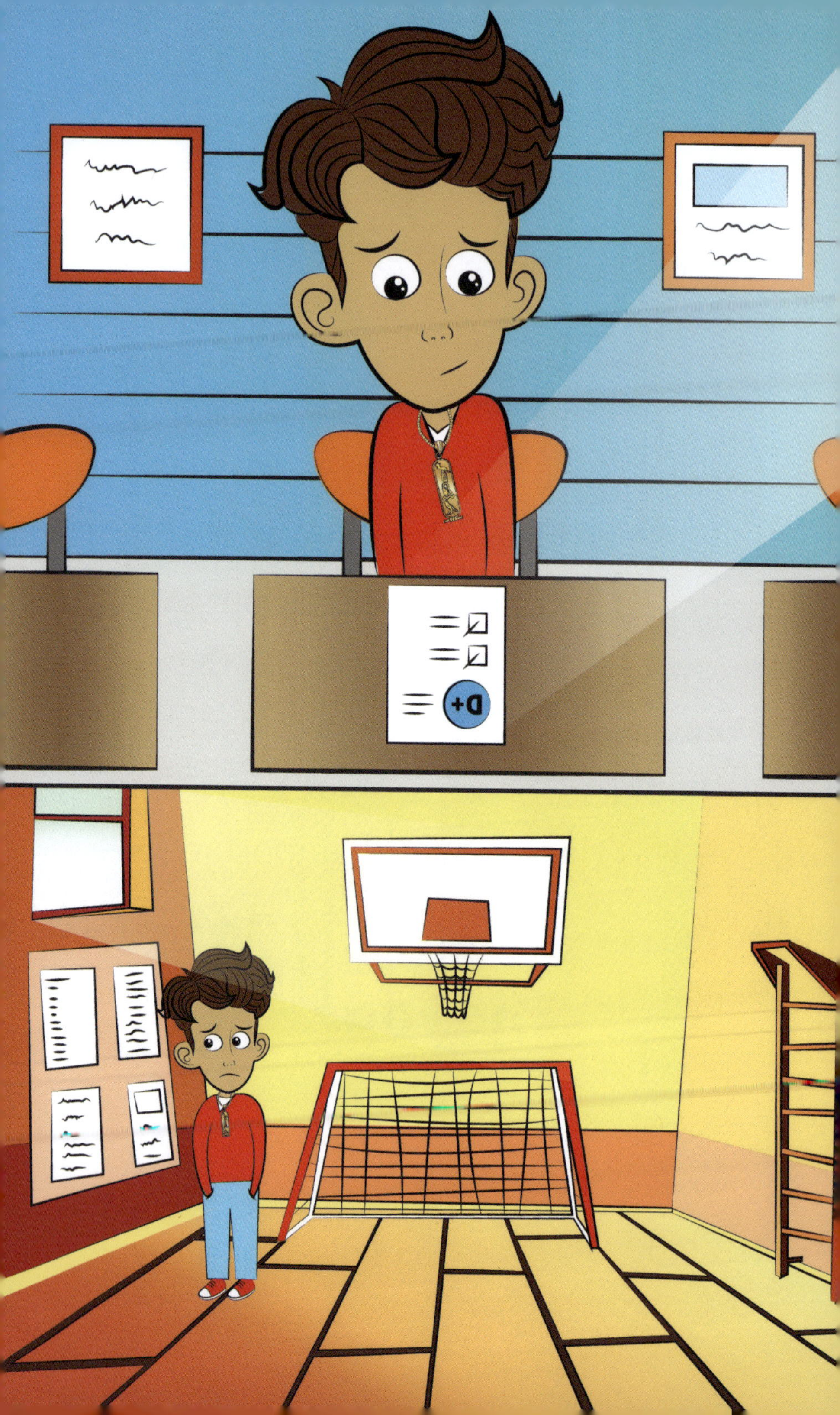

Shine Bright

Tom squirmed in his seat, waiting for his English teacher to put his quiz on his desk. Silence cloaked the room, other than the rustle of people turning over their papers, and the odd gasp as they read their score. Tom's stomach was a mess of tangled knots. Finally, Miss Thomson slapped his test on the desk in front of him. He glanced up at her with a hopeful glimmer in his eyes. But she only pressed her lips together in disappointment.

His heart leapt into his throat as he

flipped it over and saw his low score. Lower than he'd expected.

After class, he slumped into the hallway, a lump in the back of his throat. But he remembered they had announced the baseball team. He hastened through the hallways that smelled of feet and cleaning products, all the way to the gym. But his name was not on the baseball team. He had not made it.

"I just feel like I'm not having enough shining moments," he said to Omar, one of his best friends, at lunch. "I keep hitting speed bumps and it's making me lose my con idence."

Omar furrowed his brows and nodded, tossing two fries into his mouth.

"It sounds like you've had a lot of bad luck."

Omar made his way back to his next class with a bitter sadness in his heart. So, he texted his friends and rounded them up at their spot beneath the oak tree by the football field.

"What opportunities can we suggest to Tom? Opportunities that will make him shine?" Omar said, glancing around the circle. The wind made the tree rustle above them.

"He could enter the art competition!" Anastasia said. "Tom is an amazing artist and paints awesome pictures of space."

"Or tryout for the soccer team," Simon said. "He's a great striker during P.E."

The friends came up with more ideas and wrote them down on a piece of paper. They also wrote Tom a note, reminding him he was a shining friend.

"What are you guys doing?" Tom said as he walked past, heading towards the bus.

"We made this for you," Omar said, handing him the paper.

Tom scanned it and tears of joy sprang to his eyes. "You are the best friends ever," he said, flashing them a watery grin.

"We want you to shine, Tom!" Anastasia said. Then she pointed to the painting he had rolled up, poking from his backpack. "Is that

№ 1 Том

your painting from class? You should enter it into the art competition. I admired it today and even the teacher commented on it!"

A week later, Anastasia scurried up to Tom in the hallway as he shut his locker. Her cheeks were pink and flushed with excitement. "Guess what?"

"What? Did Mr. Cromwell set his desk on fire again?" Tom asked with a chuckle.

She giggled. "No. Your painting won a place in the competition! It's pinned to the wall of fame in the art hall."

His eyes widened. "Show me!"

He followed her through the hallways and his violet solar system painting hung high on the wall. Tom pulled his friend into a hug. "I can't thank you and the others enough. Your kindness and support nudged me to enter when I didn't think I could shine."

The End.

Turn the page!

Bonus Activity

At Home Word Wall

Create your own word wall at home to learn vocabulary words that you do not know!

Why should you do this?

Not only is this word wall great for building vocabulary, spelling and comprehension skills, but you can also get creative and decorate the word wall however you want by adding a fun border and background. Keep your handy word wall up year-round or use it for homework or when studying for spelling tests.

At Home Word Wall

Here's how you do it...

Look at the vocabulary list and place a star or emoji next to the words that you know. Grab a notebook or printer paper to write down all the words without the mark next to them. This will become your list for the Word Wall.

You can use fun alphabet prompts and blank cutouts. It looks better if you use printer paper, markers or colored pencils, and the secret ingredient is you as the artist!

See a word that you do not understand? Flip the page or return to the Table of Contents to find the vocabulary list and follow the steps below.

At Home Word Wall

7 EASY STEPS
TO MAKE YOUR OWN WORD WALL

1. Grab a sheet of printer paper.
2. Fold it in half (long ways)
3. Write the word large on the outside.
4. Google the definition or meaning of that word.
5. On the inside, write the definition and draw a picture that represents the meaning of this word.

At Home Word Wall

7 EASY STEPS
TO MAKE YOUR OWN WORD WALL

6. Secure the word to a board, bathroom wall, or back of your bedroom door with tape (allowing the flap to remain open to see).

7. Repeat this for every word that you do not know.

Tip: Google "creative word wall ideas" or "at home word wall". Search "word wall" on ▶ YouTube YouTube to see cool examples.

At Home Word Wall

Ready for a challenge?

Challenge yourself to write a story using as many words from the word wall as you can. Return to the Word Wall when you are uncertain of a word's meaning and add new words as you want!

Brag about it!

This will look super cool...and you will want to brag about it! Get your family's permission to post it. Snap a photo of the Word Wall and post it with the hashtags #creativegenius #imovealot. You can always tell a friend about your Word Wall.

Vocabulary List

ancient	disinterested
autopilot	dye
beamed	erupt
bitter	expression
blobbed	fiction
blurted	flicking
cauldron	flinched
civilization	frazzled
competition	fright
cosplay	frustration
creaked	furrowed
cricket (sport)	gasp
crimson	giddy
curried	giggle
demonstrate	glance

glimmer	nudged
glistening	P.E. (Physical Education)
grim	peroxide
grin	prickling
heaved	rack
hunched	scorpion
hurled	scurried
immersive	shattered
lava	slouched
leapt	slumped
mammal	species
marveled	squirmed
mildew	strained
monarchy	tangled
mosque	thumped
non-fiction	unwind

Teacher Resource

Classroom Word Wall

Create a classroom word wall for Grades 2 to 4.

Why is this beneficial?

The activity below supports all 3 learning styles (audio, visual, and kinesthetic learning) using Bloom's Taxonomy level 2 (understanding) and level 6 (creating). Students will participate in social learning to have fun with language arts.

Need more convincing?

Edgar Dale's Cone of Experience states that after two weeks people remember "90% of what we say and do". A Classroom Word Wall is a great way to ensure a higher rate of knowledge retention for your students.

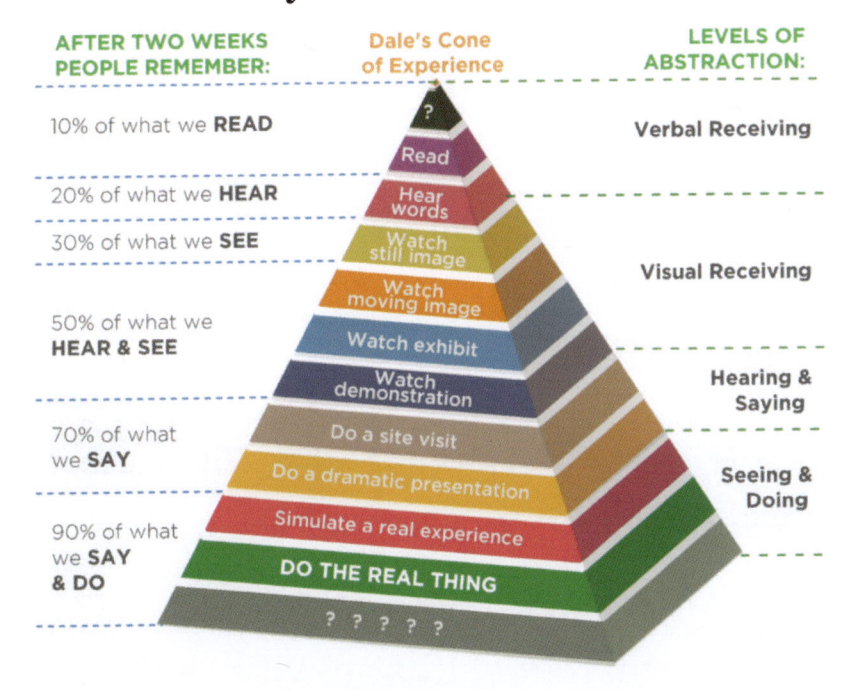

AFTER TWO WEEKS PEOPLE REMEMBER:
- 10% of what we **READ**
- 20% of what we **HEAR**
- 30% of what we **SEE**
- 50% of what we **HEAR & SEE**
- 70% of what we **SAY**
- 90% of what we **SAY & DO**

Dale's Cone of Experience:
- ?
- Read
- Hear words
- Watch still image
- Watch moving image
- Watch exhibit
- Watch demonstration
- Do a site visit
- Do a dramatic presentation
- Simulate a real experience
- DO THE REAL THING
- ? ? ? ? ?

LEVELS OF ABSTRACTION:
- Verbal Receiving
- Visual Receiving
- Hearing & Saying
- Seeing & Doing

Dale's Cone of Experience (Dale, 1969)
Flores, M. (2017, July 2). Are serious games a new approach to learn about lean? A story from Pratt & Whitney. Lean Analytics Association. Retrieved February 8, 2023, from https://lean-analytics.org/are-serious-games-a-new-approach-to-learn-about-lean-a-story-from-pratt-whitney/

WORD WALL Classroom Activity

Create a classroom word wall for grades 3, 4, and 5

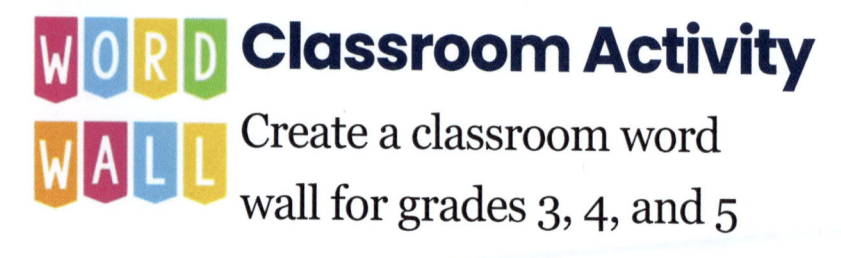

Opening Activity: The teacher will compile a list of the new words on the Whiteboard from the Vocabulary List. After assigning one word to each student, the teacher will give specific directions for this activity.

Teacher Tip: Use different colored paper to help students recognize different kinds of words.

Main Activity: Assign students, possibly working in pairs, a term to define for the class word wall. The teacher will model this activity on the whiteboard with an 8.5 x 11 sheet of paper. After folding it in half (long

Classroom Activity (continued)

ways), the teacher will write the word large on the outside. On the inside, the teacher will write the definition and draw an image or graphic that represents the meaning of this word. Then the teacher will secure the word to the board with tape (allowing the flap to remain open for viewing).

Differentiation: Facilitating personal word walls for second language learners, students of varying reading skill, students with learning disabilities, and younger learners.

Assessment: Ask students to present their word before placing it on the word wall. This is an informal way to assess learning.

Classroom Activity (continued)

Teacher Tip: New words can be added to the word wall as needed. Students can also update the definitions on their own word walls as they develop a deeper understanding of key terms.

Extend the learning: Challenge students to write a story using as many words from the word wall as they can. Encourage students to revisit the Word Wall when they are uncertain of its meaning. This will save you time in explanation and help students have ownership in their learning. Teach them to fish!

There's More!

AMAZINGLYSHERMAINE.COM

SCAN TO COLLECT THEM ALL!

COLORING BOOKS

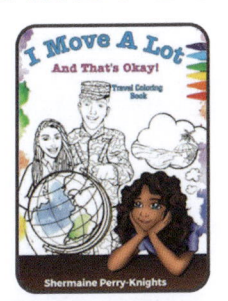

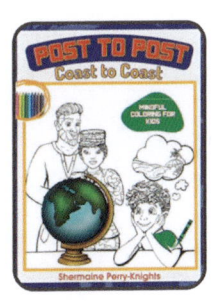

PICTURE BOOKS

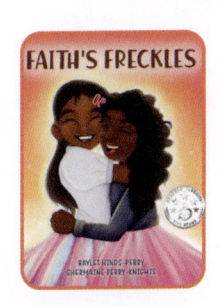

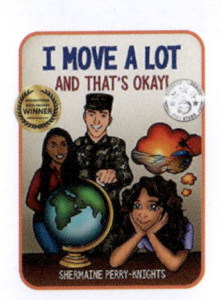

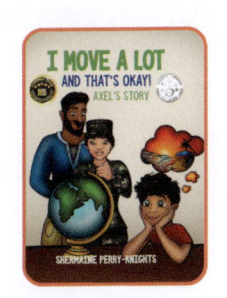

About the Author

Shermaine Perry-Knights is a TedX speaker, an award-winning facilitator and world traveler. She is a proud third-culture kid and a lifelong learner that uses her experiences in keynotes on resilience and change management. Her books and workshops make military-connected families feel "heard and seen."

Shermaine is a Certified Professional in Talent Development. She earned a bachelor's of arts in political science from Spelman College and a master's in public administration from Strayer University. Shermaine Perry-Knights has authored guided & bullet journals, youth & adult coloring books, and children's books on self-love and resilience.

Scan the QR Code to view exclusive freebies, books and merchandise. You can book Shermaine for speaking events on www.amazinglyshermaine.com.

Do you want Ms. Shermaine to read to your school?

Ask your teacher or principal to email her on this website.

amazinglyshermaine.com/contact